Starting from Themes

Transport

Starting Points for Work in Cross-curricular Mathematics

by Peter Clarke

1SBN 1 874099 54 5

Edited, typeset and designed by Fran Mosley
Cover design by Anne Gilchrist
Illustrations by Andy Martin
Printed by GPS Ltd, Watford

The BEAM Project is supported by
Islington Education

CONTENTS

INTRODUCTION

Why cross-curricular mathematics?

"In life, experiences do not come in separate packages with subject labels. As we explore the world around us and live our day-to-day lives, mathematical experiences present themselves alongside others."

Mathematics Non-Statutory Guidance (1989)

Children's mathematical experiences fall into two categories. Pure mathematics focuses on activities such as ordering numbers or investigating triangles. Mathematics which is embedded in a context arises from work on cross-curricular topics or in 'real-life problem solving' and involves problems such as finding out who is the tallest in the class.

In a balanced curriculum children need both kinds of experience. They need to learn pure mathematics in order to gain a thorough grasp of skills and facts; but it is only through using mathematics in a variety of contexts that children fully develop their understanding of relevant concepts.

Cross-curricular work in the classroom

"The application of mathematics in contexts which have relevance and interest is an important means of developing pupils' understanding and appreciation of the subject and of those contexts."

Mathematics Non-Statutory Guidance (1989)

It is important that when children work with cross-curricular mathematics the task has a clear purpose. If it does, then:

- the work has meaning for the children
- the work is motivating (because it has a purpose)
- the children can take control of the mathematics, choosing methods that suit them
- they are likely to feel confident in tackling tasks
- the context provides many clues and stimuli to support children's thinking
- they can use a common-sense approach, relating the task to past experience

- the mathematics is practical rather than abstract, and builds on children's previous experience

When planning mathematical work it is important to spend time thinking about the rationale for the tasks and to identify learning objectives for mathematics. These might consist of:

- knowledge you would like children to gain
- skills or competences you would like them to acquire
- concepts you would like them to develop
- attitudes you want to foster towards learning, and towards themselves and each other

These objectives can come from several sources: evaluation of what children need next or what they need more experience of; consideration of the statements in the national curriculum programmes of study; your school's long-term coverage requirements (how often areas need to be covered and when); attitudinal and social needs; equal opportunities considerations.

Once the list of objectives is drawn up you can use it to develop ideas for appropriate mathematical links with the class topic, or plan a 'mini-maths topic' to run alongside. The class topic may need to be researched beforehand to supplement your own knowledge. The result should be a set of activities which meets the learning objectives and provides for children working at different levels.

At the back of this book you will find a sheet for noting your learning objectives for any one activity, and how children achieve these objectives — there is space on the sheet for making assessments on up to six children. This sheet may be freely photocopied for use in school.

Assessment

Once you have identified the learning objectives, the whole process of ongoing assessment is more manageable. One option is to make a simple matrix with learning objectives listed down one side and a group of children's names along the top, and use this to record sample observations. Another option is to collect samples of work for children's portfolios which show the fulfilment or otherwise of specific aims. Your evaluation after the lesson or group of lessons can focus on how well these aims were met and can help form your next set of plans.

How this book can help

This book contains many starting points for mathematical activities on the topic of transport. These could be over-whelming without a strategy for accessing them. But planning learning objectives, as described above, offers a way in.

The starting points are cross-referenced to the national curriculum for mathematics as well as other areas. Each starting point can be developed into a wealth of learning experience, so you may want to choose just one or two, and concentrate on tasks derived from them.

Many of the starting points can provide work at two or more levels: what children learn will depend on how you present and develop the ideas, and at what level you choose to pitch them. This means that you can use the book with a wide age range, and with children of various abilities.

Choosing the tasks

One way to choose tasks is to hold a brain-storming session with the children to produce ideas which they would like to explore further. You can of course suggest your own ideas — this is an opportunity to assess which ones appeal to the children.

Next you should compare the raw ideas thrown up by brain-storming with the starting points in this book and use these to refine the children's ideas. Again, you may want to introduce some ideas of your own at this stage. It is unlikely that the children's ideas will suffice to cover the learning objectives you have in mind. You can now produce a list of questions and activities for children to explore.

Organising the tasks

Presentation

Once you have selected which starting points to work from, you will need to present them to children. You could:

- write each starting point or task on a card, and store these in a box for easy reference (either as one class set, or duplicate sets for each group)
- list the tasks in the form of a class poster

Ways of working

Ways of working vary according to your and the class's style. Some examples are:

- the children choose from a selection of tasks, in any order, either individually, in pairs or groups

- you decide which children do which task, and plan for a variety of individual, paired and group work
- a combination of these two ways, where you decide groupings, but children choose their starting points
- the whole class uses one starting point to begin with, then you support individuals or groups in the direction in which the starting point develops

It is a good idea if children spend some time working through an idea, rather than dealing with it superficially — the more involved an investigation becomes, the more learning will take place, as children make connections between findings. However, you will have to decide when enough time has been spent on any one enquiry. Some ideas could become a year's work!

Involving the children

Whichever style you adopt (and it may consist of combinations of the above), it is important that you involve the children in the tasks from the very beginning. Encourage children to suggest ways the starting points might develop and think of alternative methods for solving the problems. You can record these ideas on a flip chart so that they can be kept for later reference, to help motivate and interest children as they move through a task.

If you present just one starting point for the whole class to work on, you can start with a session where children suggest ways in which it can be developed. The children can follow up their own ideas, with awareness of what the others are doing. When the groups share their findings, children will have some understanding of each other's work.

Sharing sessions

It is common practice for children to share language work at the end of sessions, but mathematics is given this treatment less often. Children *can* present and share work in mathematics, talking about:

- what they have found out
- what they have done so far
- their plans for what to do next
- anything that went wrong and what they did about it
- the method or methods they used
- what they were pleased with
- what they thought would happen

These sessions are important. It is often

only at this point that children fully understand important mathematical concepts. They hear a variety of contexts and activities explained, and suddenly grasp all the interrelationships, as the ideas fall into place.

Your role is of course crucial, to tie all the threads together during these sessions.

Resources

If children are to take control of their mathematical learning, they need access to a wide range of resources from which they can choose what seems appropriate for the task in hand. Such resources include measuring devices, calculators and counting equipment, and paper and card of different thicknesses, sizes and types. Children also need personal jotters or piles of scrap paper on each table, so they can try things out without having to be perfect first time round.

In addition you may want to involve children in collecting together resources for the work suggested in this book. The following will be particularly useful:

— travel brochures and guides
— newspapers and magazines
— in-flight magazines
— timetables
— photographs and pictures of all varieties of transport
— maps, atlas and globe
— access to the Internet

Estimation

It is important that children are provided with opportunities to estimate and write down an approximate answer before carrying out a calculation. In *Mathematics Counts* (the report of the committee of inquiry into mathematics teaching), Dr W H Cockcroft states: 'that ability to estimate is important not only in many kinds of employment but in the ordinary activities of adult life'.

Many of the starting points included in this book provide opportunities for children to make estimations and approximations. Teachers should aim to develop in their children

- the ability to obtain, before a calculation is carried out, a 'rough answer'
- the ability to realise whether an answer is reasonable.

Starting Points for Activities

The following section makes suggestions for starting points which you and the children can build up into activities.

ON THE GROUND

MATHEMATICS	OTHER AREAS	
handling data recording	*English* • discussion	How do people travel on land? What are good methods of travelling for different conditions (snow, mountains, flat country . . .)?
handling data number	*geography* • transport	Carry out a traffic survey on the road near the school. What kind of vehicles go past the school? How many of each kind? Record your findings on a graph.
number handling data	*English* • discussion	Find out some statistics about road traffic and pollution. Discuss the pros and cons of building more roads.
measures handling data recording	*science* • forces	Try rolling a toy car down a slope. What happens as the slope gets steeper? What happens when your toy vehicle has a heavy load? Does it go as fast? Does it go as far? Record your findings.
measures handling data recording	*science* • forces	Try rolling a toy car downhill on a variety of surfaces. On which ones does it go faster? On which ones does it go further? What happens when your toy vehicle has a heavy load? Is there any difference? Record your findings.

MATHEMATICS	OTHER AREAS
maps & plans money number shape & space measures	*geography* • maps & plans *DT* • designing • making *IT* • control *art* • making

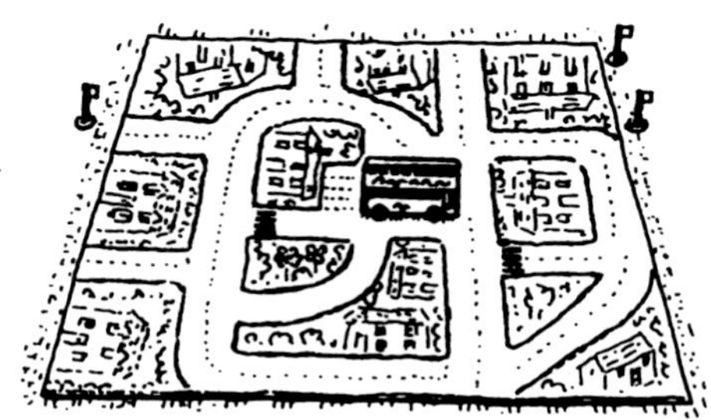

Draw a large plan of a village/town/area (like a playmat). Decide where the route will be for a toy bus. Where are your bus stops?

Draw a small map for your driver, to show where the bus has to stop.

How much does it cost to go one stop on your toy bus?

How much does it cost to go three stops?

How much does it cost to go all the way to the end of the route?

Make some tickets for your toy bus.

Devise a program for the Roamer to travel on your route.

measures
handling data

Gather together a collection of toy cars and lorries.

Which of them goes fastest with one push on the flat?

Which of them goes furthest? Record your findings.

time
recording

Mark the departure times from a bus or train timetable on a 24 hour time-line.

What do you notice?

time
recording

Look at this timetable. How long would it take to get from here to the seaside/park/your Grandma's?

ON THE GROUND

MATHEMATICS	OTHER AREAS	
measures averages		On average how far do you walk in a day? A week? A year?
measures number money	*geography* • transport	How much petrol does a family with a car use in a week? What things do you need to consider? How much would this cost a year?
shape & space		Draw and name all the shapes you can find on a bicycle, a car, a train . . .
measures problem solving	*science* • forces	How much does the teacher's car weigh? How could you work it out?

WHEELS

MATHEMATICS	OTHER AREAS	
number estimation		How many wheels are there in the car park? In the classroom? In the town?
number		How many wheels are there on 1 car, 2 cars, 3 cars, 4 cars . . ?
number sorting	*geography* • transport	Collect or draw pictures of wheeled vehicles and sort them according to how many wheels they have.
number problem solving		In the garage there are 21 wheels, all belonging to bikes or trikes. What is the highest number of vehicles there could be in there? And the lowest? Suppose there were 22 wheels?

MATHEMATICS	OTHER AREAS	
shape & space	*science* • forces	Make some different shaped wheels and attach them to a vehicle. What happens if the wheels are squares, triangles or ovals?
pattern sorting & classifying	*art* • making	Investigate the different shape of the hub-caps on car wheels. Make sketches of the different patterns. Find a way of sorting your patterns. Ask others to guess your rule for classifying the patterns.

WHEELS

MATHEMATICS | **OTHER AREAS**

measures
pattern
angle
symmetry

Measure some car wheels and carefully outline the pattern on a large sheet of paper.

Study the pattern on the hub-cap What angles can you measure?

Do the patterns have any lines of symmetry or rotational symmetry?

number

How many spokes do bicycle wheels have?

Do bigger bicycle wheels have more spokes?

measures
position
movement

science
• motion

How are bicycle wheels usually measured when you buy them in a shop?

measures
rotation
circles

Find a real or toy wheeled vehicle. How far forward do the wheels move in one complete rotation?

Measure the distance moved in a complete rotation by wheels on some other vehicles.

Measure their diameters. What do you notice?

pattern
combinations

Philippe has four colour flashes to decorate his bicycle wheel. They are red, blue, green and yellow. In how many different ways can he decorate his wheel?

What if he has five colours? Or six?

HIGH FLIERS

MATHEMATICS	OTHER AREAS	
number measures		How high does an aeroplane fly? How many times higher than the school is that?
number mass*	*English* • research skills	What is the mass of a Boeing 747? (How would you find out? Ring up an airline; look in the library; search on the Internet; look through in-flight magazines?) What would be the approximate total mass of the staff, the passengers, and their luggage? So what is the total mass of the plane and contents? How many cars does this equal?
measures	*geography* • place	How long does it take for a plane to travel to Sydney?
measures number		How fast does a passenger aircraft travel? If you could travel that fast, how long would it take you to get home after school?

* In the National Curriculum for mathematics the term 'weight' no longer appears, but has been replaced by the term 'mass'. Weight, the amount of pull something exerts, is properly measured in newtons; mass, the amount of substance, is measured in grams.

HIGH FLIERS

MATHEMATICS	OTHER AREAS	
money measures	*geography* • place	What is the total paid by the passengers travelling on Concorde from London to New York?
3D shape measures handling data	*science* • forces & motion *DT* • making	Design a range of aeroplanes made from A4 paper. Which flies furthest? Measure and record your finding. Which do you think is the best and why?
plans	*geography* • maps	Draw a map of what you think your school would look like from an air-balloon.
time measures handling data	*geography* • place *English* • research skills	Visit an airport and find out some facts about it, such as: — how many flights there are in a day — how many people use the airport each day — what jobs people have who work at the airport
shape & space	*science* • forces & motion *history* • historical knowledge	What shapes are the wings and bodies of aeroplanes? Are all aeroplanes the same in this way? How have the shapes of planes changed over time, and why?
distance money	*geography* • place *English* • research skills	In one year a businesswoman flies from London to Paris and back, to Rome and back, to New York and back and to Los Angeles and back. How many kilometres does she travel? (How could you find out?) How much would this cost?

WATER, WATER, EVERYWHERE

MATHEMATICS	OTHER AREAS	
maps & plans distance	*geography* • maps • transport	Find as many rivers and canals as you can on a map of Britain. (You may want to look at a special canal map.) Plot various routes that a canal boat could take starting from a point near where you live.

shape & space measures	*DT* • designing • making *science* • floating • forces	Design and make a canal boat to float on water. What materials will you use? What mass can it carry before it capsizes or sinks? How could you make it carry a greater mass?
shape & space measures	*science* • forces • energy	How are different boats and ships powered? Can you make a model boat that moves through the water? How will it be powered?
measures	*science* • displacement	Make two balls of plasticine the same size. Make one of them into a boat and leave the other as a ball. Put them both carefully into the water. What happens? Why is that?

WATER, WATER, EVERYWHERE

MATHEMATICS	OTHER AREAS	
handling data recording measures		Gather together a collection of toy boats. Can you sort them in different ways? Have a race. Which of your toy boats travels the furthest? Which travels the fastest? Why? Record your findings.
handling data	*geography* • place	Visit a port. Find out what the port is used for. How many people use it each day? How much cargo passes through the port in a day? Where do the ships travel to and from?
location angles coordinates	*science* • Earth	How do captains of ships know where they are? How do they know how to find a port?
money time distance maps & plans	*geography* • transport • place • maps *English* • discussion	By sea, what routes are there for crossing the English Channel/Irish Sea/North Sea? How much does each one cost? How long does each one take? Which would you use and why?

COMMUTING

MATHEMATICS	OTHER AREAS	
handling data graphs	*geography* • transport	Carry out a survey of how teachers and parents travel to work. What kind of transport is used? How many people use each kind? Record your findings on a graph.
handling data graphs	*geography* • transport	Carry out a similar survey on how children in the class gets to school.
handling data graphs percentages reasoning	*geography* • transport *English* • discussion	What are the different ways people commute to work? Where we live, what percentage of people do you think use each type of transport, approximately? Which is the most popular? Would it be the same in other parts of the country? If not, why not?
cost distance time	*geography* • transport *English* • discussion	In our area, what would be the cost of commuting to and from the same place for a week by rail, car, taxi or bus? What travelling times would be involved in each mode of transport? Which would you choose? Why?
cost distance time	*geography* • transport *English* • discussion	Which mode of transport would you choose to use if you were going to travel 10 km every day for a year? Why?

DELIVERING THE GOODS

MATHEMATICS	OTHER AREAS	
money measures	*geography* • place • maps • transport	How much does it cost to post a letter to someone in England, Europe, New Zealand? Consider the miles that each letter has to travel. Which letter is the cheapest per mile? What vehicles might the letter travel in?
money measures	*geography* • place • maps • transport	How are goods transported to different parts of the country? Does transporting goods a long distance affect their price?
measures number	*geography* • place • maps • transport *English* • discussion	Imagine a lorry load of biscuits travelling to Cornwall from a biscuit factory in Scotland. Try to work out approximately how much petrol is needed to carry the biscuits that distance. Try to think of reasons for and against transporting food from one end of Britain to the other.

MATHEMATICS	OTHER AREAS	
fractions handling data	*geography* • place	Choose 20 pieces of packaged food. What fraction of them was produced in your local area? Elsewhere in Britain? Overseas?

DELIVERING THE GOODS

MATHEMATICS	OTHER AREAS	
money number	*geography* • transport • place *English* • discussion	When goods are imported or exported these prices are affected by exchange rates. What are exchange rates? What is the exchange rate today between England and other countries? What about last week? Was it the same? If not, why not? What does all this mean for the price of imported and exported goods?
money distance time	*geography* • transport *English* • discussion	If you were to export something to Spain which mode of transport would you use? What things would you need to consider?
3D shape nets	*DT* • designing • making	Gather together a collection of items used for postage — envelopes, boxes, tubes . . . How are these designed to ensure safe delivery of posted items? Take them apart in order to find their nets. Now copy the net to make a package like it. Design and make your own packaging for an item to send through the post.
3D shape capacity	*DT* • designing • making	Use construction materials to make your own toy truck. How much can it carry? What is its capacity? Can you redesign it to carry even more goods?

HOLIDAYS

MATHEMATICS	OTHER AREAS	
maps distance	*geography* • place • maps • transport	Where in Britain have people in this class visited? Can you locate those places on a map/atlas? Approximately how far away are those places? How did the people travel there? How else could they have travelled there?
maps distance	*geography* • place • maps • transport	Where outside of Britain have people in this class visited? Can you locate those places on a map/atlas/globe? Approximately how far away are those places? How did the people travel there? How else could they have travelled there?
cost	*geography* • place • maps *English* • writing for a purpose	Choose a country and collect as much information as you can about it. Make a brochure about it. What modes of transport would you need to use to get there? How much would it cost?
maps routes cost time	*geography* • place • maps	Choose somewhere that you could visit on a class outing. How would you travel? What route would you take to get there from your school? What places would you pass through? How much would it cost? How long would it take?

HOLIDAYS

MATHEMATICS	OTHER AREAS	
maps money number	*geography* • transport • place • maps *English* • writing for a purpose	Choose a tourist attraction in your local area and collect as much information as you can about it. Make a brochure about it. Make a map to show visitors how to get to your chosen place. Is it possible to reach it by rail or coach? Now organise a visit. Does it live up to your expectations?
money distance time	*geography* • transport • place • maps *English* • discussion	Decide on a place in Britain you would like to visit. What are the different ways you could travel there? Work out how long each one would take. How much would each mode of transport cost? Which is the best way to travel and why?
maps distance time	*geography* • place • transport *English* • discussion	You are flying away on holiday. Where does your holiday start from? What time will you need to leave home to be sure to arrive in time to catch your plane? What things will you need to consider? 
3D shape capacity	*DT* • designing • making	How much English currency would you need on a holiday abroad for a week? What things would you need to consider?

OTHER IDEAS

MATHEMATICS	OTHER AREAS	
3D shape	*DT* • making	Make a model car (or boat/sledge . . .) that moves.
sorting recording	*geography* • transport	Collect together models of various modes of tranport and sort them according to how they move. Now find a different way of sorting them. Can you record your sorting?
ordering time	*history* • chronology	Collect pictures of various modes of transport and put them in order according to when they were used. Now put them in order of when they were invented.
time	*history* • chronology *English* • discussion	Make a time-line showing the major developments in transportation since 1900 (or since the wheel was invented). Discuss the pros and cons of each development.
measures	*English* • research skills	What is the largest vehicle in the world? And the smallest? Where would you go to find out this information?
classifying measures	*science* • energy • resources • health *English* • research skills • discussion	Investigate how vehicles are powered. Which kinds of power are more polluting and which are 'greener' than others. In what way?

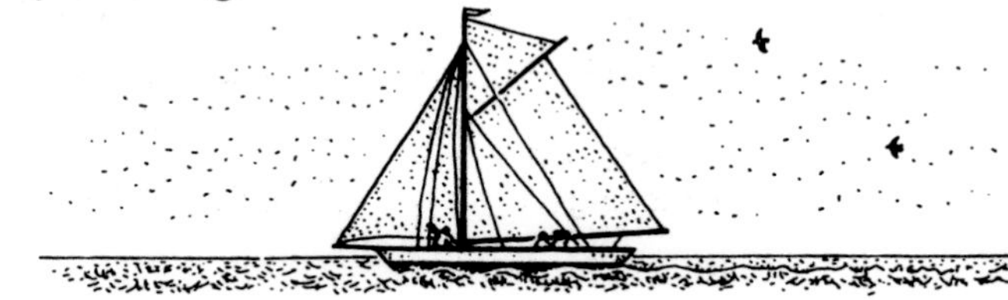

OTHER IDEAS

MATHEMATICS	OTHER AREAS	
3D shape nets	*DT* • designing • making *English* • writing for a purpose	Make a vehicle with wheels using balsa wood/Lego/card/junk boxes. Write instructions for somebody else to make a vehicle just like it. Include a template if you think that would help.
3D shape measures	*DT* • designing • making	Make any kind of toy vehicle that can carry 500 g mass. Make a toy vehicle with 500 cc capacity.
measures		Here is a pedometer. How does it work? What does it do? Use it to find how far a journey round the school is.
measures		Investigate milometers on cars.
maps time money handling data	*geography* • place • transport • maps & globes *art* • making *English* • writing for a purpose	Set up a travel agency in the home corner. Advertise the cost, travelling time and distance to certain places using various modes of transport. Set up a database with suitable information about the places people can visit. Make some adverts offering cheap flights. What prices and destinations would be realistic?
money	*English* • discussion *DT* • designing • making	When most people travel a long distance they either pay by cheque or by credit card. How do each of these operate? Design a cheque or credit card.

OTHER IDEAS

MATHEMATICS	**OTHER AREAS**	
money	*geography* • place *English* • discussion	Taxis are a mode of transport that exists all over the world. How are they different in Asia, Venice, London, New York and other parts of the world? What coins would you pay the taxi driver with if you were in Spain, Poland, Bangladesh, South Africa, the USA . . ?
money time measures	*geography* • place • transport	What are the different ways you can travel from London to Barcelona/Helsinki/Delhi/Hong Kong/Nairobi? Which is the quickest? Which is the cheapest? Which way would you choose to travel and why?
measures	*English* • discussion *geography* • place • transport	Submarines, camels, tanks, cable-cars, hovercrafts, and husky dogs are all unusual modes of transport. Where and why are they used? How are they used? How fast do they travel? What weights can they carry? What else can you find out about unusual modes of transport?

Ways of Developing Starting Points

This section shows how various teachers developed three different starting points for use in class. These can provide models for how you plan work based on starting points of your choice.

Starting point 1:

Design and make a canal boat to float on water. What materials will you use?

What mass can it carry before it capsizes or sinks?

How could you make it carry a greater mass?

The teacher developed the above starting point (taken from p.17 of this book) into activities suitable for whole class and group work in a mixed ability Year 2 class.

Learning objectives

The teacher identified these mathematics, science, and design and technology learning objectives:

- selecting mathematics to use for a task
- devising a test
- comparing strengths and other properties of different materials
- comparing efficiency of different designs
- estimating and measuring accurately standard length and weight
- finding ways of overcoming difficulties
- modifying and improving designs
- reflecting on fairness or otherwise of different tests
- working cooperatively in group
- presenting results in a clear and organised way
- contributing to discussion by offering suggestions and ideas

Introducing the work

a class discussion on ships and boats (what they carry, what they are made of, how they are designed to float and carry heavy weights)

↓

teacher presents the class with a range of available materials that could be used to make a boat and class discusses these materials

↓

in groups of three children design a boat to carry a small weight

↓

as a whole class make a decision as to the four designs which are likely to be the most successful

↓

divide into four groups to make the designs chosen

Group work

each of the four groups use available materials to design their boat

↓

each group devise a test for strength using different weights of plasticine

↓

identify weakness in design, try out improvements and retest

↓

record results

Whole class conclusion

at the end of the work bring the whole class together

↓

each group present their results to the class with an emphasis on: reflecting findings; deciding which tests worked well and why there may have been difficulties

↓

brainstorm the way forward

↓

compare each water vessel using the same test

↓

record results

Starting point 2:

Choose a tourist attraction in your local area and collect as much information as you can about it.

The teacher developed the above starting point (taken from p.23 of this book) into activities suitable for group work with a mixed ability Year 6 class. After introducing the activity the teacher then planned in detail how to help the class investigate the following:

— setting up a database with suitable information

— making a brochure

— making a map to show visitors how to get to the chosen place

— whether it is possible to reach it by rail or coach

— how much it costs to reach the tourist attraction

Learning objectives

The teacher identified these mathematics, geography and information technology learning objectives:

- selecting mathematics to use for a task
- devising a chart to collect data
- calculating with time and money
- refining approximation and estimation skills
- contributing to discussion by offering suggestions and ideas
- using measuring instruments
- obtaining information from other sources
- presenting information in a clear and organised way
- using a database
- designing a map
- making generalisations
- working cooperatively in small groups

Introducing the work

whole class discussion on tourist attractions in the local area

↓

class decide on one tourist attraction

↓

class is split into five groups and each group is told their task

Group 1 — setting up database

collect as much information as they can about the tourist attraction

↓

prioritise the information

↓

information is put onto database

Group 2 — making brochure

group are given a number of pamphlets of the tourist attraction

↓

decision is made as to what is to be included in the brochure

↓

produce brochure advertising the tourist attraction

Group 3 — making map

locate the tourist attraction on a map of the local area

↓

draw a map from the school to the tourist attraction

Group 4 — rail or coach

investigate the possibility of reaching the tourist attraction by rail and coach

↓

find out how long each mode of transport takes to get from school to the tourist attraction

Group 5 — calculating cost

investigate the cost involved in travelling by both coach and rail from the school to the tourist attraction

↓

find out cost of entrance, if any

↓

consider whether any other costs are likely and if so approximate amounts

Whole class conclusion

at the end of the work gather whole class together while each group presents the results to the class

↓

class decide which mode of transport would be the most suitable

↓

work out plans for whole trip

↓

decide on who to inform and tell them

Starting point 3:

Carry out a traffic survey; carry out a commuting survey

The above starting points, taken from pages 10 and 19, were developed by a group of teachers into activities to be used by a group of children in any of the mixed ability classes in the school.

Learning objectives

The teachers identified these mathematics and geography learning objectives:

- conducting a survey
- keeping a tally
- selecting mathematics to use for a task
- designing a chart to collate data
- constructing and interpreting statistical diagrams
- giving possible reasons of statistical findings
- liaising appropriately with parents and teachers when collecting information
- presenting results in a clear and organised way
- working cooperatively in small group

Activity for any group

discuss purpose of survey: why are they doing a survey; what information are they looking for?

↓

discuss and plan:
how to carry out survey;
what questions they need to find answers to;
how to work systematically;
how long it will take;
where and when to do it

↓

conduct survey

↓

look at and discuss results: how should these be collated; how should they be presented; is a computer database needed?

↓

collate results

↓

discuss: do results make sense; do they answer the original question?

↓

present findings in graphical and written form; discuss whether these make sense

↓

identify and discuss similarities and differences; look for patterns

↓

report findings, giving own interpretation of results

Date .... **Activity** ....

Learning objectives	Name	Name	Name	Na

Using this sheet

This chart has been designed for use with a group of about six children who are working on one of the activities in this book. It may be photocopied for use in school.

Along the top of the chart you can write up to six children's names. In the left-hand column you can write in the main learning objectives, which may be from mathematics or other curriculum areas such as language, science or technology. How each child does in meeting each objective can be written in the appropriate space below that child's name.

When writing your main learning objectives in the left-hand column you may want to leave room to make general observations about the child's work, or add other learning objectives that may have been met but that were not intended at the planning stage.

We would like to thank the following teachers and schools for their help with this book:

Mel Ahmet and Thornhill Primary School, Islington
Nirosha Attale and St Mary's CE Primary School, Islington
Catherine Clark and Prior Weston Primary School, Islington
Kim Connor and St John Evangelist Primary School, Islington
Nicky Ford and Penton Primary School, Islington
John Spooner and Rotherfield Junior School, Islington
Margaret Clark and Downhills Primary School, Tottenham
Diana Cobden, Mathematics Adviser, Dorset